YOUR KNOWLEDGE HAS VALUE

An infinite set of hybrid functions with one unique member whose verifiable zeros are to be found only on Riemann's Critical Line and nowhere else in the Critical Strip

William Fidler

Bibliographic information published by the German National Library:

The German National Library lists this publication in the National Bibliography; detailed bibliographic data are available on the Internet at http://dnb.dnb.de.

ISBN: 9783346725110
This book is also available as an ebook.

© GRIN Publishing GmbH
Nymphenburger Straße 86
80636 München

Print and binding: Books on Demand GmbH, Norderstedt, Germany
Printed on acid-free paper from responsible sources.

The present work has been carefully prepared. Nevertheless, authors and publishers do not incur liability for the correctness of information, notes, links and advice as well as any printing errors.

GRIN web shop: https://www.grin.com/document/1272674

An infinite set of hybrid functions, with one unique member whose verifiable zeros are to be found only on Riemann's Critical Line and nowhere else in the Critical Strip

W M Fidler

Abstract

We have devised a hybrid function, denoted by H_a, (where a, is a real constant), which consists of the linear combination of a novel form of the Riemann zeta function and the abscissa of any point in the complex plane. These functions comprise an infinite set, for the value and algebraic sign of the constant is unconstrained.

Amongst these functions $H_{1/2}$ is unique, in that, the magnitude of its value at the intersection of any Dirichlet line [1] with Riemann's Critical Line [2] is shown to be absolutely zero and that there are no other zeros of this function anywhere else in the Critical Strip. There may be other zeros of this function elsewhere in the complex plane but we argue that this can never be proved; this is a feature of any other of the H_a whose zeros can be posited to exist at the intersection of a vertical line passing through any abscissa of choice with a Dirichlet line but, can never be shown to be exactly zero, since this would require that the Dirichlet alternating eta series associated with the real part of these H_a be summed to infinity.

It follows from the above that, for the function, $H_{1/2}$ Riemann's hypothesis is verified.

2

Contents

Introduction

Other than the interest in the zeta function in its own right as a series it could be said that such interest increased exponentially when Euler showed that the zeta function was directly related to the sum over the prime numbers. The zeta function also arose in Riemann's 1859 [2] paper which was concerned with the determination of the number of primes in a given range, and where he extended the zeta function to the whole of the complex plane by analytic continuation. Riemann showed that all of the zeros of this extended zeta function in the negative half of the complex plane lay at the locations $x = -2, -4, -6$, etc, whilst all of the zeros in the positive half lay within a region extending from $x = 0$ to $x = 1$. Moreover, it seemed that all of the zeros lay on the vertical axis of symmetry, which Riemann called the Critical Line, but he could not prove this. Since then, the proof that all of the zeros in the positive half of the complex plane lie on this line of symmetry has acquired a life of its own and is considered to be one of the most important unsolved problems in Mathematics.

There are versions of the zeta function whose only zeros lie on the lines of symmetry passing through $x = -2, -4, -6$ etc, and this is the subject of [9], but we show here that there is a hybrid function, $H_{1/2}$ whose only provable zeros lie on the Critical Line. In addition, on the basis of numerical calculations, it is posited that there are other hybrid functions which pass through the same points on the line as those of $H_{1/2}$ but this can only be regarded as a probability for it can never be verified analytically.

Analysis

Although having been presented before [3], much of the following is considered essential to the exposition of the subject of this work and is repeated without apology.

The Riemann zeta function, $\zeta(s)$ is an extension to the series:

$$\zeta(s) = \frac{1}{1^s} + \frac{1}{2^s} + \frac{1}{3^s} + \frac{1}{4^s} + \frac{1}{5^s} + - - - - \text{---------------} \quad (1).$$

Here, the real number exponent is replaced with a complex number, $s = x + i\,y$.

It should be noted that we use Riemann's notation for the complex number but the normal mathematical notation for its real and imaginary parts.

Under the same constraint as above we write the Dirichlet eta function, $\eta(s)$ as :

$$\eta(s) = \frac{1}{1^s} - \frac{1}{2^s} + \frac{1}{3^s} - \frac{1}{4^s} + \frac{1}{5^s} - \text{------------------------} \quad (2).$$

From equations (1) and (2) we get: $\qquad\qquad\qquad\qquad\qquad\qquad \zeta(s) -$

$$\eta(s) = 2^{1-s}\left[\frac{1}{1^s} + \frac{1}{2^s} + \frac{1}{3^s} + \frac{1}{4^s} + - - - - \right] = 2^{1-s}\,\zeta(s).$$

For reasons which will become apparent later in the analysis we write the above as:

$$-\eta(s) = (2^{1-s} - 1)\,\zeta(s) \text{----------------------------}(3).$$

Now, Riemann's functional equation is: $\zeta(s) = 2^s \pi^{s-1} \sin\left(\pi\,s/2\right) \Gamma(1-s)\zeta(s-1)$.

It then follows that we may write the Dirichlet functional equation in terms of the Riemann functional equation

Hence, $\eta(s) = (1 - 2^{1-s}) 2^s \pi^{s-1} \sin\left(\pi\,s/2\right) \Gamma(1-s)\zeta(s-1)$. \text{----------------} (4)

We now seek a solution to equation (1) when $\zeta(s) = 0$.

Equation (1) is written out in extenso :

$$\zeta(s) = \frac{1}{1^{x+iy}} + \frac{1}{2^{x+iy}} + \frac{1}{3^{x+iy}} + \frac{1}{4^{x+iy}} + \frac{1}{5^{x+iy}} + \text{------} \quad \text{----------------} \quad (5).$$

For the sake of illustration consider the second term of the above.

5

i.e. $\quad 1/2^{x+iy} = e^{-iy\,ln2}/2^x$, which by Euler's theorem may be written=

$1/2^x[cos(y\,ln2) - i\,sin(y\,ln2)]$

Hence, we may collect terms in equation (5) and write:

$$\zeta(s) = \sum_{k=1}^{k=\infty} 1/k^x \cos(y\,lnk) - i \sum_{k=1}^{k=\infty} 1/k^x \sin(y\,lnk) \text{ -----------------------} (6).$$

If we set $y\,lnk = k\,\pi$ it then follows that all of the imaginary terms disappear. This leaves the

real part to be given by: $\sum_{k=1}^{k=\infty} 1/k^x \cos(k\,\pi)$.

Hence, the real part of equation (6) becomes : $- 1/1^x + 1/2^x - 1/3^x + 1/4^x - 1/5^x$ -----

This may be written: $- [1/1^x - 1/2^x + 1/3^x - 1/4^x + 1/5^x \text{ ------}]$.

But, this is equal to $-\eta(x)$.

If the real part of (6) is to disappear then from equation (3) we have:

$-\eta(x) = (2^{1-x} - 1)\,\zeta(x) = 0.$

Either the first term on the RHS disappears or the second.

The first term will vanish if $x = 1$, but this would reduce equation (1) to the Harmonic series which is known to diverge, albeit slowly, to infinity. The product above would then be of the form: $0 \cdot \infty$, which is indeterminate. It then follows that we must take $\zeta(x)$ to be zero.

We now return to the functional forms for ζ and η.

Again, Riemann's functional equation is: $\zeta(s) = 2^s \pi^{s-1} \sin(\pi s/2)\,\Gamma(1 - s)\zeta(s - 1)$.

Now, the sine term will vanish if we set $s = -2n$, where n is a real integer.

It then follows, that setting $s = -2n$ will yield the following result: $\zeta(-2n) = 0$, and hence, from equation (3), $\eta(-2n)$ will also vanish. It is important to emphasize that the sine term will vanish if, and only if, n is an integer.

This procedure outlined is, of course, that employed in generating the so-called 'trivial zeros' of the Riemann zeta function. Further, from the functional equation we see that whatever ζ is evaluated to on the left of $\mathbf{s} = \frac{1}{2}$ is determined by its evaluation on the same point reflected across $\mathbf{s} = \frac{1}{2}$. This, together with the analytic continuation of η provides the ability to compute ζ anywhere in the complex plane. In addition, since $\zeta(s)$ has no zeros to the right of $\mathbf{Re(s)} = 1$, then the functional equation predicts that there are no other non-trivial zeros to the left of $\mathbf{Re(s)} = 0$ and hence all of the non-trivial zeros lie within the Critical Strip.

The function $F(x + i\,y_k)$

In previous work [1] we established the existence of lines in the complex plane, which we named, Dirichlet lines. This was accomplished by writing the function $F(x + i\,y_k)$ in the form of a zeta function with $y_k = \frac{k\pi}{\ln k}$ throughout the infinite range of the function. Although upon perusal of the equation written out in extenso it seemed that the equation was a zeta function of the Riemann form, in that all of the exponents of the terms in the denominators were complex numbers, it could be shown that all of the terms in the series were real and summed to the negative of Dirichlet's alternating η function.

We now consider a zeta function, $\overline{F}(x + i\,y_k)$, where $y_k = \frac{k\pi}{\ln k}$, with the exception of a single term. We consider $\overline{F}$ to be a Riemann zeta function, albeit novel, for it fulfils the necessary criteria, in that it has the form of the Riemann zeta function, the sum of which may be a complex number. The following, and continuation thereof are discussed at length in [9].

We now write out this function in the form of that shown in equation (5), but with a single exceptional term as shown: i.e. $\overline{F}\;(x + i\,y_k) = \frac{1}{1^{x+i\pi/\ln 1}} + \frac{1}{2^{x+i2\pi/\ln 2}} +$

$\frac{1}{3^{x+i3\pi/\ln 3}} + \cdots \frac{1}{(j-1)^{x+i\,(j-1)\,\pi/\ln(j-1)}} + \frac{1}{j^{x+iy_j}} +$

$\frac{1}{(j+1)^{x+\,i\,(j+1)\pi/\ln(j+1)}} + \ldots\ldots\ldots \qquad \ldots\ldots\ldots\ldots (7).$

Consider the term, $\frac{1}{j^{x+iy_j}}$.   This

may be written in the form $\frac{1}{j^x}\left[\cos(y_j \ln j) - i\sin(y_j \ln j)\right]$ ------------- (8).

We are free to stipulate how the function, $\overline{F}$ behaves, and, as shown above we have chosen that the **j**th term of the series may be written as shown in equation (7), which expression we now write in the compact form:

7

$$\bar{F}(x + i\,y_k) = \sum_{k=1}^{k=j-1} 1/_{k^{x+i\,k\pi}/_{\ln k}} + 1/_{j^{x+i\,y_j}} + \sum_{k=j+1}^{k=\infty} 1/_{k^{x+i\,k\pi}/_{\ln k}} \quad \text{------ (9).}$$

Now, the addition of the two summations in equation (9) will yield the negative of the Dirichlet alternating η function without the quantity, $1/_{j^x}\cos(y_j \ln j)$, whilst all of the imaginary functions in the summations will have disappeared with the exception of the remaining imaginary term in the equation i.e. $-i\,1/_{j^x}\sin(y_j \ln j)$. If we now set $\quad y_j = {}^{j\pi}/_{\ln j}$ then this remaining imaginary term will also disappear and the sum given by equation (9) will be real and equal to the negative of the η function.

The hybrid function

We now set down, by fiat, a hybrid function, H_a given by

$$H_a = G[(a - x) + iy_k] + x \quad \text{-------------------- (10).}$$

This expression may be taken to represent an infinite set of equations if we place no restriction on the value or algebraic sign of the constant, **a.** We stipulate that **G** is a zeta function, and so we may then rewrite this equation in the same form as that which produced equation (7); i.e

$$H_a = 1/_{1^{(a-x)+i^\pi}/_{\ln 1}} + 1/_{2^{(a-x)+i2\pi}/_{\ln 2}} + 1/_{3^{(a-x)+i3\pi}/_{\ln 3}} +$$

$$\dots 1/_{(j-1)^{(a-x)+i\,(j-1)\,\pi}/_{\ln(j-1)}} + 1/_{j^{(a-x)+iy_j}} +$$

$$1/_{(j+1)^{(a-x)+\,i\,(j+1)\pi}/_{\ln(j+1)}} + \dots\dots\dots + x \quad \dots\dots\dots\dots \text{(7).}$$

Let us set $y_j = {}^{j\pi}/_{\ln j}$. As we have shown before, the function **G** will now become real, the imaginary terms having vanished and, the real part given by the negative of Dirichlet's alternating eta function; hence, H_a becomes; $H_a = -\eta(a - x) + x$ -------------- (11).

Setting a = ½ we see that at x = ½, $H_{1/2}$ becomes, $-\eta(0) + 1/2$.

Now, $\eta(0)$ is Grandi's series, and it was shown in [4] that this series has the value of **1/2**, if, and only if, the number of terms in the series is infinite.

8

It follows that, the zeros of $H_{1/2}$ are all located at the intersections of Dirichlet lines with Riemann's Critical Line and so we see that Riemann's hypothesis is verified for this function.

As before, if we insert the value of **y** into the iterative equation derived in [5] and find that the result for **j** is not an integer, then we conclude that the product **y ln j** is not compatible with an integer multiple of π; hence, not only does the real term not equal that required to complete the summation for the Dirichlet function, but even, if by some act of serendipity the summation is realized, there will still be a single imaginary term, for,- i **sin(y ln j)** does not vanish.

The function H_a now becomes a non-zero complex number.

We now investigate the behavior of the function $H_{1/2}$ on Riemann's Critical Line when, $y_j \neq$ $j\pi/\ln j$.

We concentrate on the Dirichlet cell (see [1]) in the Critical Strip centered on **x =1/2,** as shown in Fig1 below.

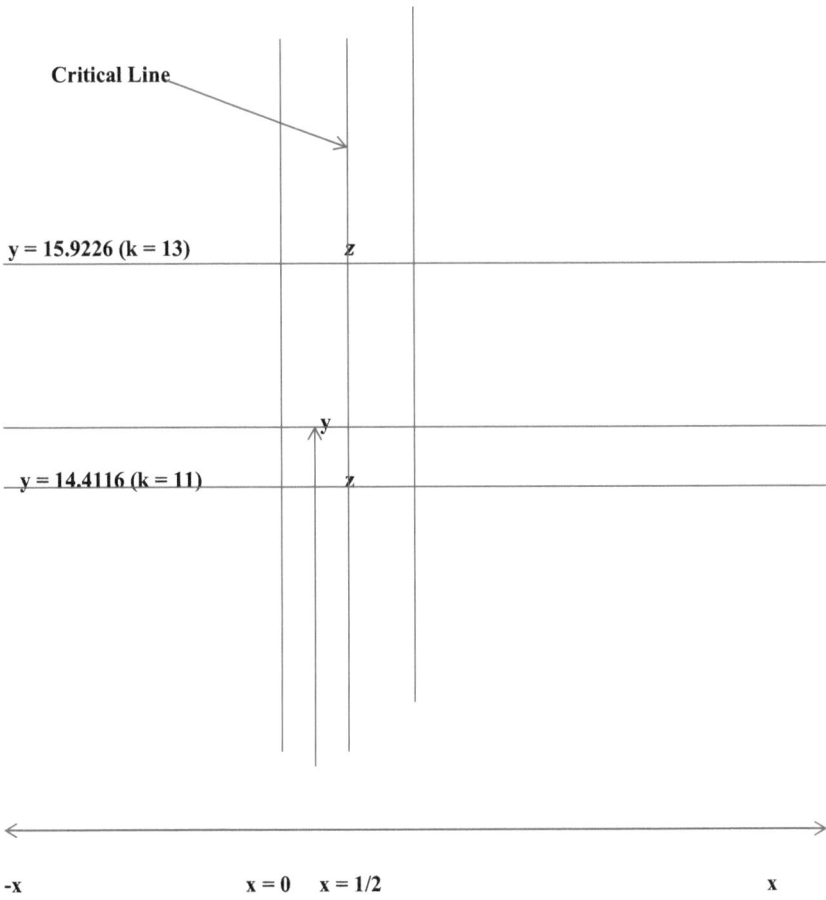

Fig1

Along the lines marked **k = 11** and **k = 13** we have set $y_j = \dfrac{j\pi}{\ln j}$ and hence the value of $H_{1/2}$ at the points marked '**z**' on the diagram is zero. If however we choose a point on the Critical Line which lies within the cell at a height, **y** as shown, then we may demonstrate that the value of $H_{1/2}$ at this point is not zero, indeed, it is a complex number.

10

To illustrate, we propose that $y = 14.823$ (a pure guess). Using the iterative equation developed in [5] it is shown that this lies between the Dirichlet lines with parameters 11 and 12, respectively. We now take $j = 11$.

Grandi's series , $S = 1 -1 +1 -1 +1 -1 + \dots\dots\dots\dots\infty$. We see that the odd-numbered terms are positive, whilst even-numbered terms are negative. However, it should be noted that here, the real part of G is the negative of Dirichlet's alternating series and hence the eleventh term in the real part of equation (7), which would then be -1 if $y_j = {11\pi}/{\ln 11}$ is now replaced by $\cos(y \ln j) = -0.55163$. The series is no longer Grandi's series and cannot be cancelled by $x = \frac{1}{2}$. The remaining imaginary part of equation (7) is not zero and is given by $-i \sin (y \ln j) = -i\, 0.8341$. In the preceding we should be mindful of the fact that the inclusion or exclusion of terms in infinite series is, as extensively discussed by Stewart [6], fraught with pitfalls.

We can repeat this argument anywhere else in the Critical Strip and show that $H_{1/2}$ is only equal to zero at the intersection of a Dirichlet line with the Critical Line.

It is easy to demonstrate, but impossible to verify analytically that $H_{1/2}$ is unique, for other of the hybrid functions may have zeros which all lie on Riemann's Critical Line*. We now show this as follows:

If we examine the plot of the eta function shown in Fig2, reproduced by the high accuracy $ke\,!^{+}$ Online Calculator [7] it is seen that the magnitude of the alternating function between $x = -8$ and $x = -10$ must pass through $\eta = 0.5$ in two places. Using the calculator for one of these places it can be shown that at $x = -9.98689$, $\eta = 0.5000000478$. Within the limits of the accuracy of this calculation, we can now determine another hybrid function whose zeros are located on the Critical Line, but it is stressed that this can only ever be a probability.

If we set $\bar{x} = -9.98689$ then $\bar{a} = -9.98689 + 0.5 = -9.48689$.

Hence, it is possible (but only probable) that the zeros of the above $H_{\bar{a}}$ exist and all lie on the Critical Line. Moreover, these will be in exactly the same positions as those for $H_{1/2}$.

*Whilst the other H_a, in theory, can have zeros in the same place as $H_{1/2}$ the fact that they are zeros at all can never be determined with absolute accuracy.

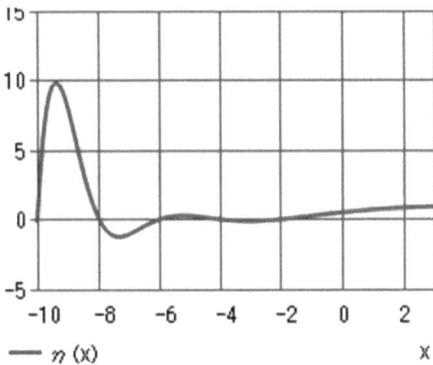

$$— \; \eta \, (x) \qquad\qquad\qquad x$$

Fig2

Since the Dirichlet alternating eta function becomes zero on passing through a trivial zero, it then follows that, since there are an infinite number of trivial zeros, then there are an infinite number of regions having the characteristics of that shown above and, bearing in mind the previous comments, it is probable, but unverifiable that there are an infinite number of hybrid functions whose zeros lie on the Critical Line all located in exactly the same places. It should be noted that if we move to a different value of **x** we can, using the procedure previously described, locate the probable zeros of other hybrid functions at the intersection of Dirichlet lines with the ordinate passing through that **x**.

Riemann's zeros and the hybrid function $H_{1/2}$.

In previous work [1] we have shown that a Riemann zero can never be located at the intersection of a Dirichlet line and the Critical Line. It then follows that the zeros, probable or otherwise, of the hybrid functions can never coincide with Riemann zeros, but, as demonstrated in [1], the Riemann zeros are contained within Dirichlet cells, and, there are an infinite number of hybrid functions whose probable zeros are located at the intersections of the upper and lower boundaries of the cell with the Critical Line.

From inspection of the following diagrams, labelled Fig2 it is easy to show that it is not possible for the negative of the Dirichlet series contained within $H_{1/2}$ to cancel with **x** except on the Critical Line. The function $H_{1/2}$ is of particular interest for it contains the only value of

Dirichlet's eta function known to the author (other than zero) that may be obtained, with exactitude without calculation, from another series, viz. Grandi's series.

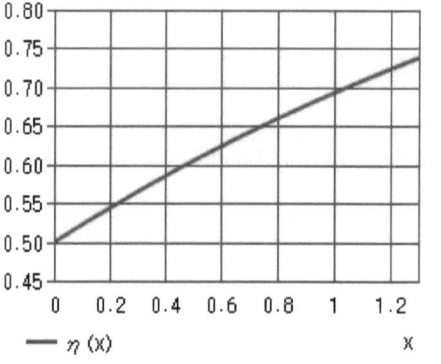
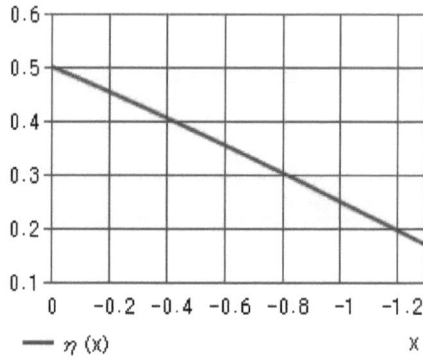

Fig3

It must be borne in mind that if we choose an **x** then we seek a value of $\eta(a - x)$ which is equal to this choice and whose negative will cancel with **x**. Inspection of the above graphs shows that cancellation of the real part of $H_{1/2}$ in the Critical Strip can only occur on the Critical Line (**x** = ½), which, of course means that any of the $H_{1/2}$ which do not lie on the Critical Line are not zero.

Further, it is concluded that since the Riemann zeros are contained within Dirichlet cells then these zeros are bounded from above and below by the zeros of the hybrid function, $H_{1/2}$, and, of course the other H_a-functions that may be obtained by the procedure outlined on p10, although again, none of the **a** for these functions can be determined exactly for this would require that their associated Dirichlet series be summed to infinity. Indeed, we may, in the light of the analysis on p10, state that, it is probable that the Riemann zeros are bounded by an infinite number of the zeros of hybrid functions which all pass through the same points as the $H_{1/2}$, but, they share the same characteristic as Riemann's zeros, which, in some instances, for the imaginary parts, have been calculated to one thousand places of decimals [8] and have not been adduced with exactitude; moreover, for these hybrid functions there are no other zeros

13

elsewhere in the Critical Strip, for, as demonstrated on p10 the 'a' for these functions it can never be determined to a degree that would constitute mathematical proof.

The trivial zeros and the hybrid function

The posited zeros of the hybrid function may be placed at the same locations as the zeros of the function, $\overline{F}$; As an illustration consider the first trivial zero at x = -2.

We require a value of $\eta(a - x) = -2$.

By examining Fig3 it may be seen that the first $\eta(\overline{x})$ that has this value lies just beyond $\overline{x}$ = -10. In addition it is seen that there is another value of Dirichlet's function which lies just before $\overline{x}$ = -12. Using the calculator [7] we obtain: $\eta(\overline{x}) = -1.999972582$ at $\overline{x} = -10.0487079$. The corresponding value of $\overline{a}$ is then, $-10.0487079 - 2 = -12.0487079$.

The prime purpose of performing this calculation is simply to emphasise again that we can never be sure that a hybrid function does have a parameter, $\overline{a}$ such that it has a zero at the intersection of any Dirichlet line with the ordinate passing through any **x**. Of course, this is with the exception of the intersections that the hybrid function with parameter **1/2** makes with Riemann's Critical Line. We cannot stress strongly enough that, for a hybrid function with a parameter other than **1/2**, we cannot evaluate the magnitude of Dirichlet's alternating function with absolute precision except at **x = ½.**

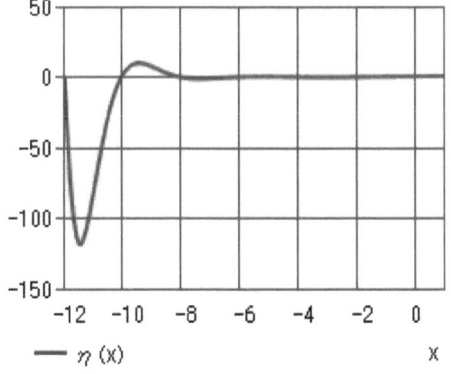

— η (x)   X

Fig4

14

Further, we can always find the probable zero of a hybrid function, for all that is required is that it lies on a Dirichlet line and $\eta(a - x) = x$. The following graphs of Dirichlet's alternating eta function versus x show that beyond $x = -2$ the function alternates with increasing amplitude about the abscissa and hence we can, in principle (but only in principle) always find an appropriate η and its corresponding parameter, **a.**

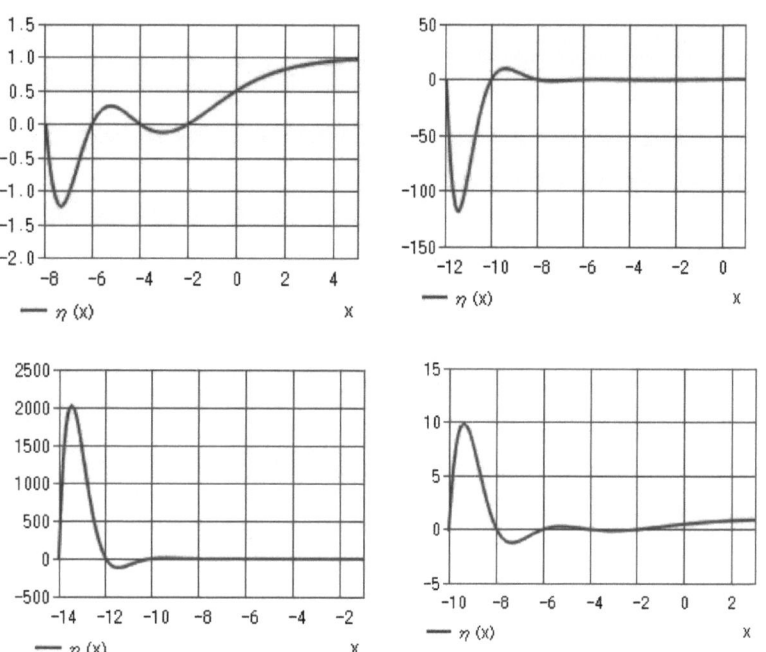

Fig5

The Critical Points

We define a Critical Point as the intersection of a Dirichlet line and Riemann's Critical Line; hence, these points have coordinates, $\left(\frac{1}{2}, \frac{k\pi}{\ln k}\right)$. It is posited that, within the Critical Strip it is only at these points that a zero value of the hybrid function may be found and, moreover, the only hybrid function which can be shown to be exactly zero there, or indeed, anywhere else in the complex plane, is, $H_{1/2}$.

Each of the Critical Points is directly associated with a counting number and so there are an infinite number of Critical Points. An interesting corollary arises from the analysis, for, the ordinate of each Critical Point on the Critical Line is given by the formula, $y = \frac{k\pi}{\ln k}$, and, for $k = 1$ the location of the corresponding point is at infinity. It then follows that the counting number associated with this point is not a prime number, and so there is not a prime number at infinity in this scheme.

We arrive at the same conclusion expressed in [5], namely that since there is a counting number associated with the only hybrid function whose value can be established to be truly zero (we exclude the other hybrid functions for the reason alluded to previously) then some of these zeros will be directly associated with prime numbers whilst others will not, and, the disposition of the prime numbers within the range of the counting number cannot be attributed to the associated disposition of the zeros, for they are one in the same.

Discussion

Much further discussion is superfluous, for the substance of this work has been laid out in great detail.

However, it is worthy of note that the hybrid function $H_{1/2}$, amongst all of the other members of the infinite set is special, in that it's value (of zero) at the intersection of a Dirichlet line with Riemann's Critical Line can be determined with absolute precision. We cannot make the same claim for any of the others members of the set for they are associated with actual summing to infinity and so their values cannot be determined with true exactitude.

In common with the function $\overline{F}$ we can never determine the y-coordinate of any intersection of a Dirichlet line with any ordinate axis for, as mentioned in previous work by the author, we do not possess an exact value for the circular constant, π.

It is worthy of note that we have made extensive use of the fact that a particular function, with only three exceptions, can never be calculated to a degree that would constitute mathematical certainty and it is somewhat ironic that we have achieved the objective of the work by constantly examining failure.

We consider that the hybrid function introduced here is a veritable cornucopia of functions, the location of the probable zeros of which may be determined at points along Dirichlet lines and thus is a subject worthy of further study.

References

[1]. The assigning of values of the prime number counting function to Bernhard
Riemann's zeros. The concept of Dirichlet lines in the complex plane
W M Fidler GRIN Verlag Cat no. v1168625, January 2022.

[2]. On the number of primes less than a given magnitude
G F B Riemann
The Monthly Notices of the Berlin Academy, Nov 1859.

[3]. On the Riemann Hypothesis
W M Fidler GRIN Verlag ISBN 9783346388575, December 2020.

[4]. On the emergence of Grandi's series with sum 0.5 as a member of an infinite set of
convergent infinite series.
W M Fidler GRIN Verlag Cat No. v386034, March 2017.

[5]. The irrelevance of the location of Riemann's zeros to the disposition of the prime
numbers.
W M Fidler GRIN Verlag Cat No. v1191150, March 2022.

[6]. Concepts of Modern Mathematics.
Ian Stewart
Dover Publications Inc, New York, 1995, Chapt 16, pp229-234.

[7]. ke!$^+$ Online Calculator.
ke!san.casio.com/exec/system/1449456065.

[8] Tables of the zeros of Riemann's zeta function.
A Odlyzko
www.dtc.umn.edu/~odlyzko/zeta-tables/index.html

[9] The verification of a Riemann hypothesis in the negative half of the complex plane.
WM Fidler
GRIN Verlag ISBN9783346698537 September 2022.

YOUR KNOWLEDGE HAS VALUE

- We will publish your bachelor's and master's thesis, essays and papers

- Your own eBook and book - sold worldwide in all relevant shops

- Earn money with each sale

Upload your text at www.GRIN.com
and publish for free